# Oxygen Delivery Systems

# And

# Mechanical Ventilation

## Made easy for House Officers

Mounir Fertikh, MD

Printed in Victoria, Canada



**Note for Librarians:** a cataloguing record for this book that includes Dewey Classification and US Library of Congress numbers is available from the National Library of Canada. The complete cataloguing record can be obtained from the National Library's online database at: www.nlc-bnc.ca/amicus/index-e.html

ISBN 1-4120-3119-2

TRAFFORD

**This book was published on-demand in cooperation with Trafford Publishing.**
On-demand publishing is a unique process and service of making a book available for retail sale to the public taking advantage of on-demand manufacturing and Internet marketing. On-demand publishing includes promotions, retail sales, manufacturing, order fulfilment, accounting and collecting royalties on behalf of the author.

Suite 6E, 2333 Government St., Victoria, B.C. V8T 4P4, CANADA
Phone 250-383-6864 Toll-free 1-888-232-4444 (Canada & US)
Fax 250-383-6804 E-mail sales@trafford.com
Web site www.trafford.com TRAFFORD PUBLISHING IS A DIVISION OF TRAFFORD HOLDINGS LTD.
Trafford Catalogue #04-0946 www.trafford.com/robots/04-0946.html
10 9 8 7 6 5 4 3 2

Dedicated to,

**Nassim Fertikh**

**Note:** This work was created outside the scope of the author's employment.
This manual represents the author's personal views and professional experience, and not necessarily those of any organization or institution.

# Acknowledgments

I wish to express my gratitude to my family, particularly my brother Dr. Djamil Fertikh for his support.

My deepest appreciation to Dr. Norman Braslow for his suggestions, guidance and encouragement. I have always enjoyed working with him.

Thanks to Dr. Naushad Shaikh as his friendship has been invaluable to me.

Thanks to Dr. Justin Sebastian for his recommendations and advice.

I gratefully acknowledge Dr. Islam Tasbirul for his thoughtful comments and suggestions.

For submitting corrections, thanks to Marlene Hamilton and Joseph Nyame, Directors of Respiratory Care.

Due to the lack of space, many deserved acknowledgments could not be made individually.

Most importantly, I sincerely hope that this manual will be useful to all house officers and students during their training in the ICU.

Mounir Fertikh, MD

**Note to the reader**

This pocket book should be used as an educational aid only and not as definite reference.
Although the informations in this book have been carefully reviewed for correctness, neither the author nor the copyright owner nor the publisher can accept any legal responsibility for any errors or omissions that may be made.
The purpose of this manual is to introduce the house officers to the basics of oxygen delivery systems and mechanical ventilation.

The reader is urged to seek further informations by consulting the suggested readings and to confirm that the informations comply with current legislation and standards of practice.

# Table of Contents

# 1. History

**Joseph Priestley** an English chemist (1733-1804) and **Karl Wilhelm Scheele** a Swedish chemist (1742-1786) discovered the oxygen molecule

**Antoine-Laurent Lavoisier** (1743-1794) a French chemist, elucidated the role of oxygen in respiration

**René Theophile Hyacinthe Laennec** (1781-1826) a French physician, inventor of the stethoscope in 1819

**John Hutchinson** (1811-1861) inventor of the spirometry in 1846, London

**Röentgen, Wilhelm Conrad** a German physicist, inventor of the X-ray in 1895

**Dr. Philip Drinker** inventor of the iron lung in 1928

**Godfrey Hounsfield** a British engineer and **Allan Cormack** a physicist, invented the computed tomography (CT) in 1972

**Paul C Lauterbur** a Professor of Chemistry at the State University of New York at Sony Brook and **Peter Mansfield** a Professor of Physics, both played the major role in developing the Magnetic Resonance Imaging (MRI)

## 2. Respiratory Physiology (Figure 1)

The respiratory system can be divided into:

**Upper airways:** Nose, nasal cavity, sinuses, larynx and trachea.

**Lower airways:** Bronchi, bronchioles and alveoli.
Between the trachea and the alveolar sacs, the airways divide multiple times, increasing the total cross-sectional area of the airways. The alveoli are surrounded by a capillary network.

**The respiratory membrane,** where gas exchange occurs is made of: The alveolar cells (the type I cells form the alveolar wall, the type II cells produce surfactant, a lipid-type substance that reduces surface tension within the alveoli) (1), the interstitial space (2), the capillary basement membrane (3) and the endothelial cells (4).

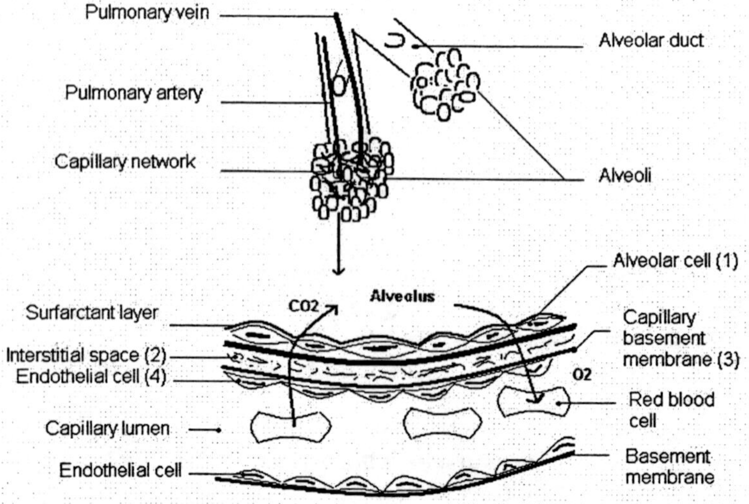


**Figure 1**
Gas exchange at the alveolar-capillary membrane.

**Mechanics of Respiration:**

Inspiration is an active process during which there is contraction of the inspiratory muscles (diaphragm, external intercostal muscles, pectoralis muscles, scalene and sterno-cleidomastoid muscles) creating a negative intrathoracic pressure, allowing air to flow into the airways.

Expiration is a passive process during which the lung and chest wall return to their resting positions. The expiratory muscles (internal intercostal muscles, rectus abdominis, transverses abdominis, internal and external obliques) are only used during voluntary expiration or during exercise.

**Control of Breathing:**

The cerebral cortex controls the voluntary breathing
The pons and medulla control the automatic breathing

The increase in PCO2 stimulates the ventilation through chemoreceptors located in the medulla, the carotid and aortic bodies.

Also, decrease in PaO2 stimulates chemoreceptors located in the carotid and aortic bodies, leading to increased ventilation.

**Breathing Patterns:**

Costal vs. Diaphragmatic

Costal breathing is a shallow breathing due to contractions of the external intercostal muscles.
Diaphragmatic breathing is a deeper breathing due to contractions of the diaphragm muscle.

**Lung Volumes and Capacities:**

Spirometry is a simple test used to measure lung function by making the patient breathe in and out through a spirometer. The results are compared to a standardized set of values based on patient's age, sex, height and weight.

All the following can be directly measured by the spirometry, except the residual volume which has to be calculated: (Figure 2)

- Tidal Volume (VT) is the volume of air that moves in and out of the lung on normal inspiration or expiration (normal ~ 500 ml)
- Inspiratory Reserve Volume (IRV) is the volume of air that can be inspired with a maximal effort after a normal inspiration (normal ~ 3.1L)
- Expiratory Reserve Volume (ERV) is the volume of air that can be expired with a maximal effort after a normal expiration (normal ~ 1.2L)
- Residual Volume (RV=TLC-VC) is the volume of air left after a maximal expiratory effort (normal ~ 1.2L)
- Vital Capacity (VC) is the largest amount of air that can be expired following a maximal inspiratory effort (normal ~ 4.8 L)
- FEV1 is the forced expiratory volume in 1 second
- Total Lung Capacity (TLC) is the total amount of air contained within maximally inflated lungs (normal ~ 6 L)
- Functional Residual Capacity (FRC) is the volume of air left at the end of normal expiration (normal ~ 2.2 L)

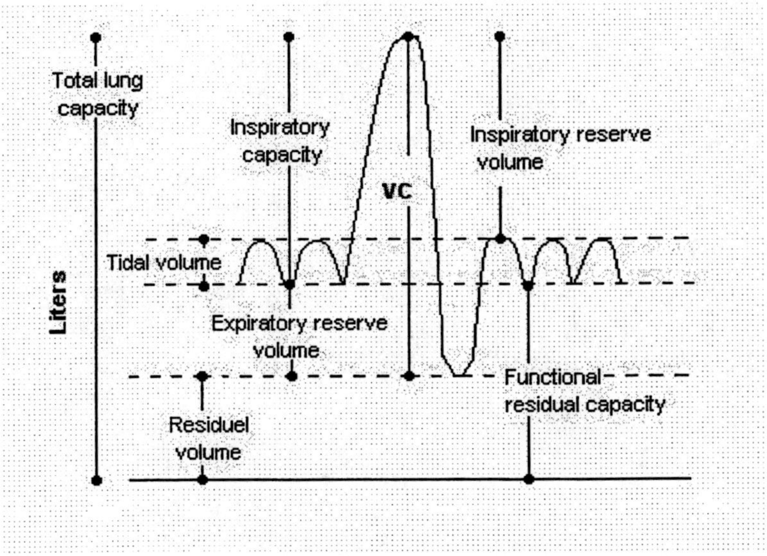


**Figure 2**
Lung volumes and capacities. VC, vital capacity.

**Table 1**

Abnormal pulmonary function tests:
Comparison between restrictive and obstructive diseases

| | Diseases | |
|---|---|---|
| | **Restrictive** | **Obstructive** |
| **FVC** | decreased | decreased |
| **FEV1** | decreased | decreased |
| **FEV1/FVC** | normal or increased | normal or decreased |
| **TLC** | decreased | normal or increased |
| **RV** | normal or decreased | increased |

**Table 2**

Abnormal pulmonary function tests:
Comparison between asthma and emphysema

| | Obstructive Diseases | |
|---|---|---|
| | **Asthma** | **Emphysema** |
| **FVC** | decreased | decreased |
| **FEV1** | decreased | decreased |
| **FEV1/FVC** | decreased | decreased |
| **TLC** | normal or increased | normal or increased |
| **RV** | normal or increased | normal or increased |
| **DLCO** | normal or increased | decreased |
| **Bronchodilator response** | Yes | No |

Peak flow meter: (Figure 3)

The peak flow meter is a portable device mostly used to monitor patients with moderate to severe asthma. It tells the patients how well their asthma is controlled.

The patient takes in a deep breath, and then makes a maximal expiratory effort through the peak flow meter with the lips closed tightly around the mouth piece.
The procedure is repeated three times and the highest number is recorded as well as the peak flow zones:

- **Red zone**: Danger zone. Seek medical help immediately
- **Yellow zone**: Warning zone. The asthma may flare. Take additional measures
- **Green zone**: Indicates that the asthma is well controlled

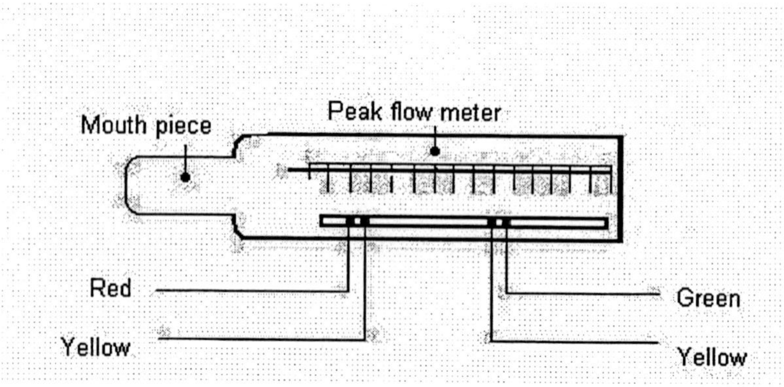


**Figure 3**
Peak flow meter.

## 3. Lung Examination

- Inspection
- Palpation
- Percussion
- Auscultation

**Inspection:**

Assess the following:

- The patient's position (lying down, sitting-up, leaning forward or resting the hands on the knees)
- Pattern of breathing
- Breathing through pursed lips (increase airway resistance)
- The ability to speak
- Abnormal sounds (eg, stridor, wheezes)
- Skin coloration, especially around the lips
- Nicotine stains on the nails and clubbing
- The shape and symmetry of the thorax (eg, barrel chest, pectus excavatum)
- The symmetry of the chest wall mouvement
- Presence of scars
- The use of accessory muscles
- The spine and scapula, looking for deformities (eg, kyphosis, scoliosis)

**Palpation:**

- Remember to rub your hands, and position them on the patient's back with thumbs pointed towards the spine

- Start from the top and work your way down while asking the patient to say "Ninety-Nine" as you proceed

- Feel for the vibratory sensations transmitted from the lung (tactile fremitus) which are increased in case of consolidation (eg, pneumonia, tumor), or decreased (eg, pleural effusion)

- Palpate the position of the trachea to rule out shift

**Percussion:**

Percussion can help identify disorders that affect patient's breathing such as: Lobar consolidation, pneumothorax and pleural effusion

- Normal lung: Resonant note
- Consolidation: Dullness
- Pneumothorax: Hyper-resonant note
- Pleural effusion: Dullness

**Auscultation:**

Please! Do not perform auscultation through clothing.

- Ask the patient to take deep breaths through the mouth
- Start from the top and work your way down, while listening over one area
- Compare the same area on the opposite side

Listen for:

- Breath sounds (normal, increased, decreased or absent)
- Presence of rhonchi, wheezes, crackles or rub

# 4. Respiratory Failure

There are two basic types of respiratory failure:

- Hypoxic
- Hypercapnic

**Mechanisms of hypoxemia:**

1. Ventilation/Perfusion (V/Q) mismatch is the most common
2. Right-to-left shunt: Cannot be corrected with 100% oxygen
3. Diffusion impairment (eg, fibrosis, granuloma, interstitial or alveolar edema)
4. Hypoventilation (eg, drug abuse, anesthesia)
5. Decrease $FiO_2$

There is desaturated blood in the arterial circulation.
The primary mechanism in 4 and 5 is a low oxygen concentration in the alveoli (low $PAO_2$).

Hypoxemia can be assessed using the A-a gradient formula:

**A-a gradient** = $P(A-a)O_2 = PAO_2 - PaO_2$

- $PAO_2$ = [$FiO_2 \times (PB - PH_2O) - PaCO_2/0.8$. It is obtained from the alveolar gas equation
- $PaO_2$ is obtained from the ABG

At sea level:
$FiO_2$ on room air = 21%
PB = Barometric pressure = 760 mm Hg
$PH_2O$ = Partial pressure of water = 47 mm Hg at 37°C

**A-a gradient** = [FiO2 x (PB - PH2O) - PaO2/0.8] - PaO2
= [FiO2 x (760 - PH2O) - (PaCO2/0.8)] - PaO2
= [150 - (PaCO2/0.8)] - PaO2

- Normally, it is less than 10 mm Hg on room air (21% oxygen)
- The larger the gradient, the more serious the respiratory compromise
- It increases with age: normal ≤ age/4 + 2.5 (on room air)

**Evaluating a patient with hypoxemia:**

If there is documented hypoxia on the ABG, calculate the A-a gradient;

**A**. If the A-a gradient is within normal value, the hypoxemia is most likely due to decrease in PAO2. The most common etiologies include:

- CNS (central nervous system) depression, drug effects
- High altitude (decrease PAO2 secondary to decrease PB)
- Disorders of neural conduction (eg, Guillain-Barré, amyotrophic lateral sclerosis)
- Muscular weakness (eg, polymyositis, muscular dystrophy)
- Rib fracture, flail chest, kyphoscoliosis
- Upper airway obstruction

**B.** If the A-a gradient is elevated; there is ventilation/perfusion (V/Q) mismatch or right-to-left shunt. To differentiate between the two, administer supplemental oxygen:

- No improvement: There is right-to-left shunt (eg, intra-cardiac right-to-left shunt, pulmonary shunt, pulmonary AVMs)

- Improvement of the PaO2: There is ventilation/perfusion (V/Q) mismatch (eg, COPD, parenchymal diseases, and pulmonary vascular diseases) or diffusion impairment (eg, fibrosis, granulomas, interstitial or alveolar edema)

| |
|---|
| Be careful when administering oxygen in chronically, hypoxic patients. Too much oxygen can precipitate suppression of the hypoxic drive with worsening of respiratory acidosis. |

**Mechanisms of hypercapnia:**

1. Hypoventilation (eg, drug abuse, anesthesia)
2. Increase $CO_2$ production (eg, fever, sepsis, thyrotoxicosis, excess carbohydrate load, over feeding)
3. Very severe parenchymal lung disease impairing $CO_2$ elimination

Hypercapnia can be assessed by the minute ventilation formula:

$$VE = VT \times f = (VA + VD) \times f$$

Minute ventilation (VE) is the volume of air that moves in or out of the lungs per minute.

Alveolar ventilation (VA) is the breathed volume of air per minute that reaches the alveoli and takes part in gas exchanges.

Dead space ventilation (VD) is that part of minute ventilation that does not take part in gas exchanges.

VT is the tidal volume, nl = [500 ml]
VD is the anatomic dead space, nl = [150 ml]
f is the frequency, nl = [12 breaths/min]

There is hypercapnia with decreased minute ventilation (decreased VT, low f, increased VD)

**Oxygen saturation curve:** (Figure 4)

The hemoglobin functions as an oxygen carrier in the human body. It binds to the oxygen at the alveolar capillary bed, travels in the circulatory system in the red blood cells, and releases oxygen to the tissues.

There is an abrupt increase in the slope of the curve: The hemoglobin molecule develops higher oxygen affinity as soon as few oxygen molecules bind to it, this is due to a phenomenon known as cooperativity. The hemoglobin binds faster to the oxygen molecule as oxygenation proceeds.

Relation between the partial pressure of oxygen (x axis) and the oxygen saturation (y axis): When hemoglobin saturation is more than 90% (the plateau of the curve), additional oxygen will not enhance delivery significantly, because the SaO2% cannot exceed 100%, it will simply increase the concentration of oxygen in the circulating blood. Thus large changes in PaO2 may occur with little change in SaO2.

PaO2 = Partial pressure of oxygen in the arterial blood
SaO2 = Arterial oxygen saturation
SpO2 = Functional SaO2 as depicted by the pulse oximeter

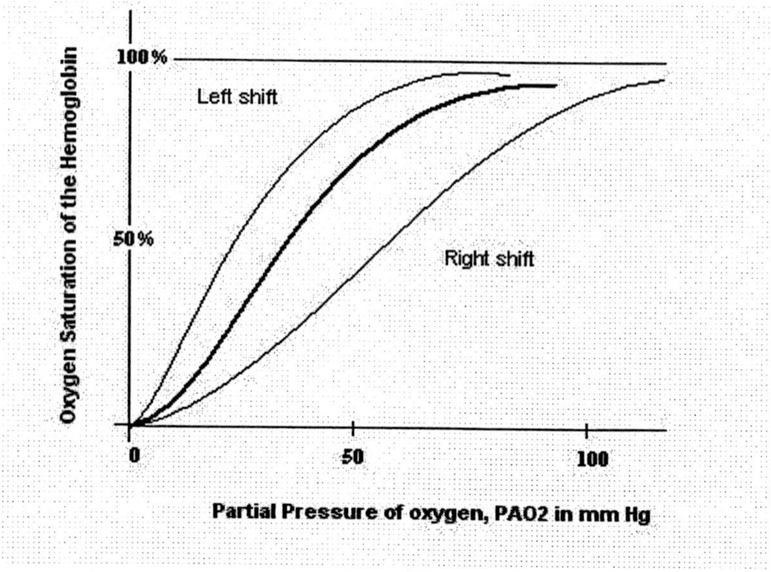


**Figure 4**
Oxyhemoglobin dissociation curve.

Shift of the curve to the left: Helps maintaining oxygen saturation in the arterial blood, with decreased oxygen release at the tissue level (increasing Hb affinity for oxygen).

- Increased pH, decreased $PaCO_2$ (eg, alkalosis)
- Decreased body temperature
- Decreased 2,3-DPG (eg, hypothyroidism, hypophosphatemia, chronic acidosis, transfusion of stored bank blood)

Shift of the curve to the right: Facilitates releasing oxygen in the peripheral tissues (decreasing Hb affinity for oxygen).

- Decreased pH, increased $PaCO_2$ (eg, acidosis)
- Increased body temperature
- Increased 2,3-DPG (eg, hyperthyroidism, exercise, residence at high altitude, anemia)

## 5. Oxygen Delivery Systems

**Nasal Cannula:**

Formed by 2 soft prongs that arise from oxygen supply tubing

- Low flow oxygen [0.5-5.0 L/min]
- FiO2 = [25%-45%], and depends on the patient's respiratory rate and tidal volume
- With each additional liter of flow, there is increase of the FiO2 by 4%
- Advantages: Low in cost, readily available, easy to set up, and well tolerated. The patient can eat, drink and talk, there is better mobility and satisfactory compliance
- Disadvantages: There is no accurate FiO2, it can be easily displaced, and in case of high flow (more than 4 liters), it can cause irritability, and drying of the nasal mucosa

    1. Flow 0 liter per minute: 21% of FiO2 (room air)
    2. Flow 1 liter per minute: 25%
    3. Flow 2 liters per minute: 29%
    4. Flow 3 liters per minute: 33%
    5. Flow 4 liters per minute: 37%
    6. Flow 5 liters per minute: 41%
    7. Flow 6 liters per minute: 45%

**Simple Mask:** (Figure 5)

Designed to fit over the patient's nose and mouth and is secured around the patient's head by an elastic band.

- Low flow oxygen, 5 to 6 liters
- Can deliver up to 50% of FiO2
- The body of the mask serves as a reservoir for the oxygen and the exhaled carbon dioxide
- The holes on each side of the mask serve as room-air entrainment ports and exit for the exhaled gases
- Advantages: Light, can be connected to small-volume nebulizers driven by oxygen and deliver aerosol therapy
- Disadvantages: Full airway enclosure (it covers the nose and mouth), skin irritation, the patient cannot eat, or drink

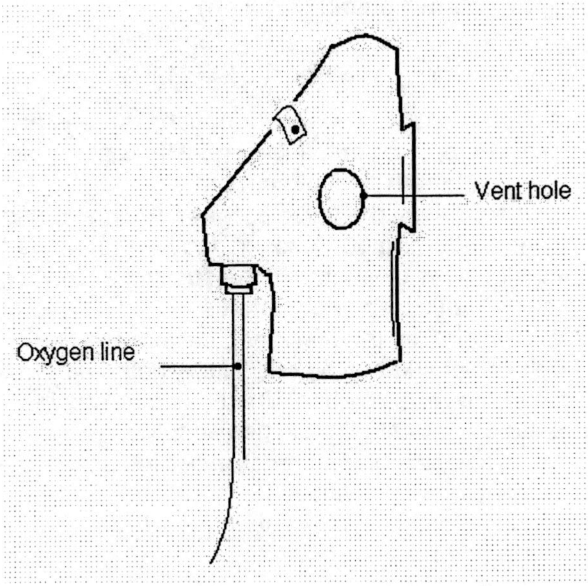


**Figure 5**
Simple face mask.

**The Non-Rebreather mask:** (Figure 6)

- Can deliver up to [90%-100%] of FiO2
- It has an attached reservoir bag that serves as a source of 100% oxygen and three one-way flaps to allow exhaled breath out but no room air in
- On inspiration, the valve between the mask and the bag opens, allowing 100% oxygen to flow to the patient. The mask valves close due to the sucking action created by the patient's inspiratory effort
- On exhalation, the mask-bag valve closes and the mask valves open letting air out. The exhaled air does not enter the reservoir bag. There is no mixing of the exhaled gas with the oxygen supply
- Therefore, the patient receives a high concentration of oxygen with each breath
- It is recommended to take one mask-valve off, so the patient can draw room air in, if the oxygen source is disconnected
- The oxygen flow should be set to keep the reservoir bag at least half inflated during patient's inhalation

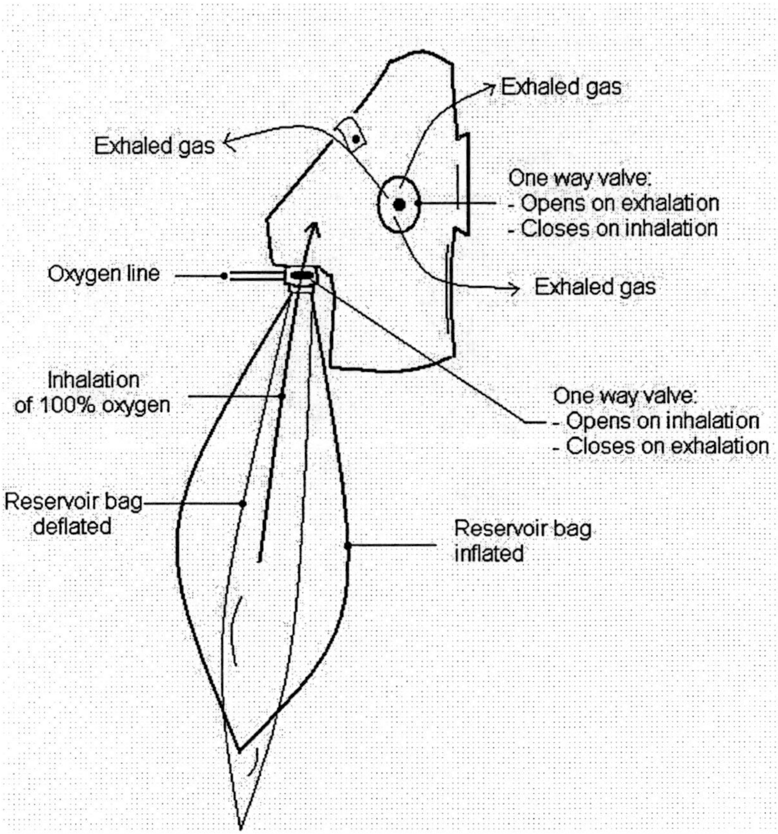


**Figure 6**
Non-Rebreather mask.

**The Partial Non-Rebreather mask:** (Figure 7)

- Here, the flap on the top of the reservoir bag has been removed, so part of the patient's exhaled air mixes with the 100% oxygen
- It has an attached valveless reservoir bag that serves as a source of oxygen and two one-way flaps allowing exhaled breath out but no room air in
- It can deliver up to [60%-70%] of $FiO_2$
- The oxygen flow should be set to keep the reservoir bag at least half inflated during patient's inhalation
- Note that in figure 7, the one-way flap located on the mask itself has been removed

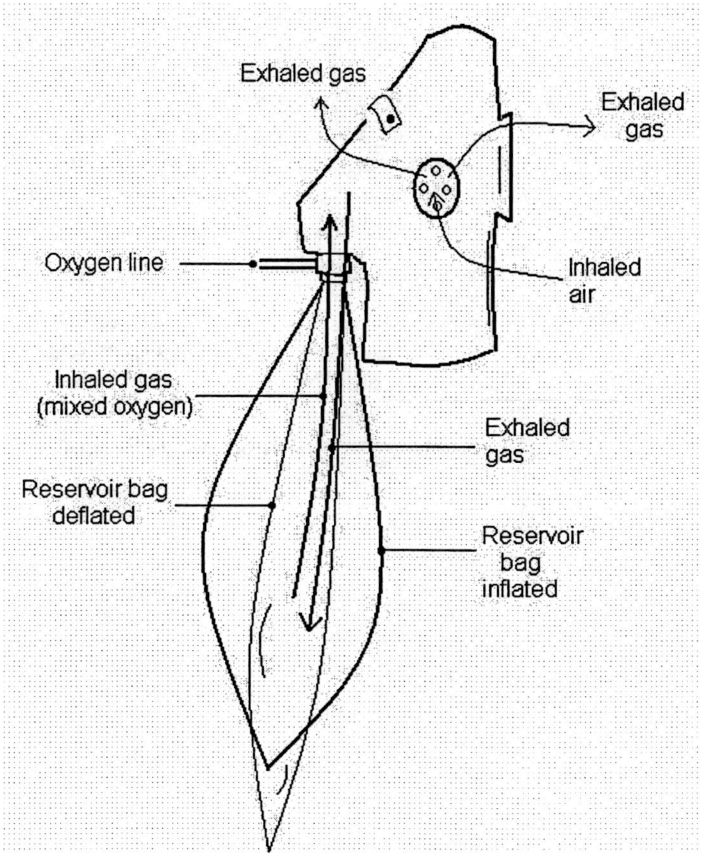


**Figure 7**
Partial Non-Rebreather mask.

**Venturi mask:** (Figure 8)

It is a mask connected to an oxygen source by wide bore tubing. The connector has interchangeable adaptors allowing the patient to receive a specific concentration of oxygen. By changing the size of the aperture and the flow of oxygen, the concentration can be adjusted to a specified amount.

- The adaptor comes in different sizes and different colors
- There is creation of a vacuum effect at the entrance of the adaptor caused by the rapid moving gas. The faster the gas flow, the greater the vacuum effect. Also the smaller the orifice the faster the flow. The high speed oxygen molecules entrain air molecules (Bernoulli's Principle)
- It is mostly used in COPD

| **Approximate oxygen concentration delivered (%)** | **Color of the adaptor** |
|---|---|
| 24 | Blue |
| 28 | Yellow |
| 31 | White |
| 35 | Green |
| 40 | Pink |
| 50 | Orange |

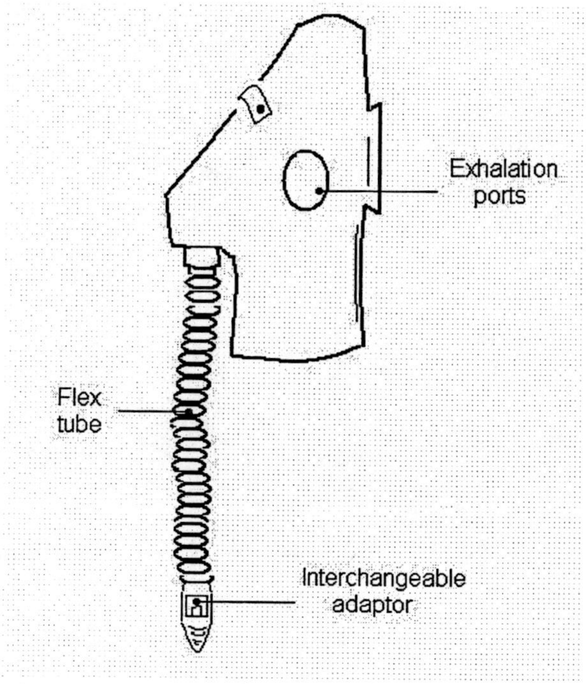


**Figure 8**
Venturi mask.

## 6. CPAP: Continuous Positive Airway Pressure

- It is a non-invasive mode of ventilation
- A tight-fitting mask that delivers a positive pressure throughout the ventilatory cycle at a specified FiO2
- The pressure is generally set at [5-10] cm H2O
- It keeps the alveoli open by maintaining a positive pressure <u>at end-expiration</u>
- There is increase of the functional residual capacity (FRC) which is the amount of air left in the lung after normal expiration
- The patient has to be awake, spontaneously breathing, hemodynamically stable and cooperative
- **Indications:** Patient with acute respiratory failure refusing intubation, hypoxic respiratory failure (eg, severe pneumonia, COPD, asthma exacerbation, lung CA, cardiogenic pulmonary edema), hypercapnic respiratory failure (eg, sleep apnea, COPD)
- **Contra-indications:** Respiratory arrest, excessive secretions, patient unable to protect the airways, agitated, non cooperative. Avoid in patients with emphysema, since they already have high FRC. Using CPAP may cause alveolar rupture
- **Advantages:** It is a non-invasive mode of ventilation, reduces the need for intubation and helps in keeping the alveoli open at end-expiration, hence improving gas exchange
- **Disadvantages:** Not very well tolerated because of claustrophobia, also can cause nasal bridge erythema or ulceration

## 7. BiPAP: Bilevel Positive Airway Pressure

- The patient has to be awake, spontaneously breathing, hemodynamically stable and cooperative
- There is a set inspiratory pressure (up to 20 cm $H2O$) and a set expiratory pressure (up to 20 cm $H2O$)
- **Contra-indications**: Respiratory arrest, excessive secretions, patient unable to protect the airways, agitated, non cooperative
- **Advantages**: It is a non-invasive mode of ventilation, reduces the need for intubation
- **Disadvantages:** Not very well tolerated (claustrophobia), also can cause nasal bridge erythema or ulceration

**How to use it:**

Start by choosing an arbitrary number,

Example: BiPAP 10/5

- 10 is the IPAP: Inspiratory pressure

- 5 is the EPAP or CPAP: Expiratory pressure

- Titrate oxygen to keep oxygen saturation > 90-92%

- Check ABG after 30 mins and assess whether pressures need to be adjusted

- If the patient is deteriorating despite of BiPAP, consider intubation

- Consult Pulmonary/Critical Care Medicine

## 8. Ventilator

The ventilator is a machine designed to support or replace normal ventilatory lung function.
All mechanical ventilators have a mechanism of controlling their cycle, which means switching from the inspiratory phase to the expiratory phase:

- **Time cycled ventilation:** Tells you that the ventilator switch from inspiration to expiration after a preset interval. It delivers a breath (inspiration) which is terminated based upon a set value for inspiration time

- **Volume cycled ventilation:** The ventilator delivers a breath after a preset tidal volume has been delivered to the patient

- **Pressure cycled ventilation:** The ventilator delivers a breath after a set pressure is reached

**Common Modes of Mechanical Ventilation:**

CMV: Continuous Mandatory Ventilation

- All breaths are mandatory, no spontaneous breaths are allowed
- All breaths are delivered by the ventilator at a preset frequency, volume, pressure and FiO2
- Used mostly in anesthesia, when the patient is heavily sedated

A/C: Assist/Control

- It is the most common initial mode used in patient with respiratory failure
- It is a CMV mode with **assisted** mechanical ventilation
- Mandatory breaths are delivered at a preset frequency, volume, pressure and FiO2
- Between the ventilator-initiated breaths, the patient can trigger the machine and receive an **assisted** breath at the volume, pressure and FiO2 set on the ventilator
- **Controls** (or provides) ventilations when patient is unable to breathe
- Advantages: There is less work of breathing and a guaranteed back up rate
- Disadvantages: Dysynchrony between the patient and the ventilator

The difference between CMV mode and AC mode is that during the AC mode, the patient can also trigger a breath. (Figure 9)

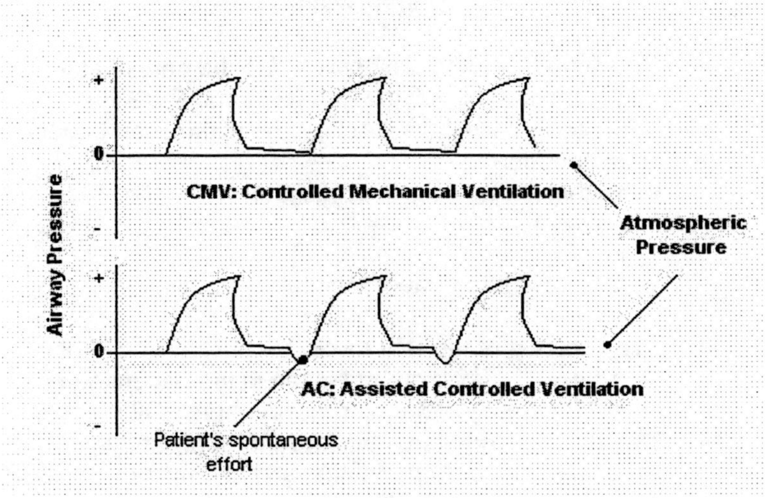


**Figure 9**
Pressure waveforms showing the difference between the CMV mode (top) and the AC mode (bottom).

IMV: Intermittent Mandatory Ventilation

- **Mandatory** breaths are delivered to the patient at a preset frequency, volume, pressure and FiO2
- Between the ventilator breaths, the patient can breathe spontaneously
- Here, there is no assisted mechanical breaths

SIMV: Synchronized Intermittent Mandatory Ventilation

- It is a version of the IMV mode
- **Mandatory** breaths are delivered to the patient at a preset frequency, volume, pressure and FiO2
- The ventilator creates a timing window around the scheduled mandatory breaths, and attempts to deliver a breath in concert with the patient's inspiratory effort
- If the patient initiates an inspiration, the mandatory breath is **synchronized** with the patient's breath
- If the patient does not initiate a breath, the ventilator delivers the mandatory breath at the scheduled time
- So, the **mandatory** breath can be ventilator or patient triggered
- Advantages: It is a weaning method
- Disadvantages: It increases the work of breathing and the oxygen consumption
- See figure 10

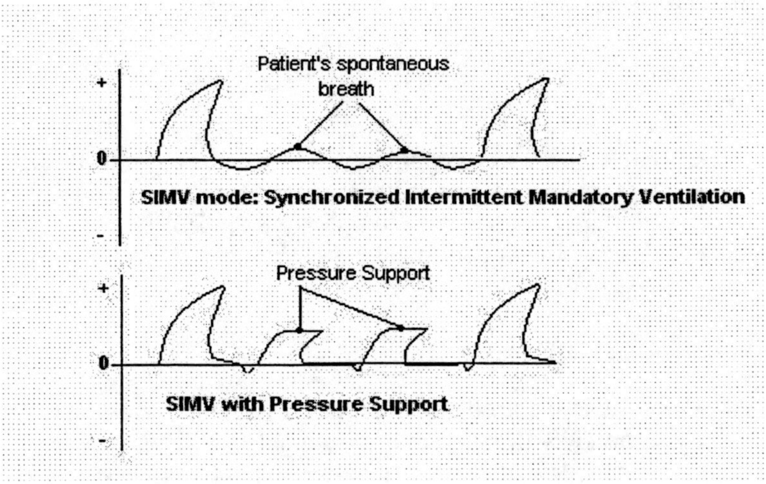


**Figure 10**
Pressure waveforms showing the SIMV mode alone (top) and the SIMV mode with pressure support (bottom).

**Ventilator Settings:**

Rate

- Start @ 8-14 breaths/min
- Set too high, it can result in respiratory alkalosis and auto-PEEP especially in patients with COPD
- Set too low, it can result in respiratory acidosis, hypoxia and hypoventilation

VT: Tidal Volume

- Volume inspired and expired with each breath
- Start @ 5-12 ml/kg of body weight
- Set too low, it can result in hypoxemia, atelectasis
- Set too high, it can result in barotrauma, decrease cardiac output and respiratory alkalosis

FiO2: Fraction of inspired Oxygen (FiO2)

- Start with 100% oxygen and titrate down to keep PaO2 > 60 mm Hg, FiO2 < 60% and SaO2 > 90% (avoid oxygen toxicity)
- Adjust by pulse oximetry or by drawing an arterial blood gas 30 mins after changing the FiO2

PEEP: Positive End-Expiratory Pressure:

- Positive pressure maintained by the ventilator at the end of expiration
- Used when the patient cannot maintain adequate $PaO_2$
- Called CPAP (Continuous Positive Airway Pressure), when the patient is breathing on his own
- Start with 5 cm $H_2O$ and titrate up with 2 cm $H_2O$ on each step. The optimal setting would be 15-20 cm $H_2O$ and is based on the patient's response
- Advantages: It keeps the alveoli open longer, and increases the $PaO_2$ without raising oxygen concentration, increases the FRC
- Disadvantages: Can cause barotrauma (pneumothorax), decreased cardiac output, hypotension, increased intracranial pressure and increased dead space

Peak Inspiratory Flow and I:E ratio:

- It is how fast the ventilator should give a breath
- It is usually set between 40-60 L/min
- Slower the flow, longer the inspiratory time; more rapid the flow, shorter the inspiratory time
- It will determine the I:E ratio, which is the time spent during inspiration versus expiration. (a conventional I:E ratio is equal to 1:2)

Peak Inspiratory Pressure (PIP):

- Preferably < 35 cm H2O
- It is the maximum pressure in the airway system at end of inspiration
- PIP = Pressure due to resistances (airways) + pressure due to compliance
- It is elevated in case of secretions, mucus plugging, kinked tube, bronchospasm, or stiff lungs

Plateau Pressure (Pplat):

- Preferably < 30-35 cm H2O
- It is a reflection of the lung compliance and is independent of airway resistance factors
- It is obtained by applying a pause at end-inspiration
- It helps in the interpretation of a high peak inspiratory pressure (PIP)
- It is **elevated** in case of parenchymal pathology, ARDS (adult respiratory distress syndrome), pulmonary edema: Decreased lung compliance
- It is **normal** in case of bronchospasm, kinked tube, secretions and mucus plugging: Increased airway resistance

Pressure Support:

- Usually set @ 10-15 cm H2O
- Used most of the time in combination with the SIMV mode or with CPAP
- It helps unloading the respiratory muscles from the imposed work of breathing caused by the endotracheal tube and the ventilator circuit
- When used with CPAP, neither the tidal volume nor the minute ventilation is guaranteed; the frequency, the tidal volume and the flow rate are determined by the patient himself
- Remember: The pressure support has relevance only for spontaneously breathing patients (eg, ventilator set on SIMV mode, CPAP)

Trigger Sensitivity:

- It allows the ventilator to **sense** and respond to the patient's respiratory effort, (the patient's negative pressure)
- Usually set at -1, -3 cm H2O
- Too sensitive can cause self-cycling and hyperventilation
- Too insensitive can increase the work of breathing
- If the sensitivity is locked out, whatever the patient does, the ventilator won't deliver any breath

VE: Minute Ventilation

The Minute Ventilation is the amount of gas delivered by the ventilator in one minute.
VE = VT x ventilator rate (liter/min). Normal = 5-7 L/min
By adjusting the tidal volume or the ventilator's rate you will maintain a desired PaCO2.

## 9. Intubation

Most common indications for intubation are:

- Patient in apnea or cardiopulmonary arrest
- Acute airway obstruction
- Patient in coma of any etiology
- Patient not able to clear secretions
- Patient with risk of aspiration, or loss of protective reflexes
- When there is increase in $PaCO2$ with associated decrease in alertness and increase fatigue
- When there is decrease in the $PaO2$ despite increase in the $FiO2$
- Massive airway bleeding
- Patient with cervical trauma would need nasotracheal intubation which will not be covered in this book

You will need:

- Gloves
- An O2 source: Mask-bag
- Pulse oximeter
- Cardiac monitor and defibrillator
- The laryngoscope, handle and blade. Check the bulb
- The ET tube [~ 7.0 mm for female, ~ 8.0 mm for male] and a fixation device
- Tape
- 10 cc syringe
- Tongue depressor
- Suction device
- Stethoscope
- Lubricating jelly (the outer surface of the tip of the ET tube should be lubricated with a water-soluble gel)

While the patient with impending respiratory failure is assisted with a bag-mask and oxygen, try to get a quick history and vital signs.

- Assess the level of distress and consciousness
- Look for foreign body inside the patient's mouth and remove it
- Examine the mouth for any denture and remove it
- Oxygenate the patient with 100% oxygen through the bag-mask
- Stand behind the patient
- Head tilted back, neck extension, jaw thrust, and chin lift, align the axes of the mouth, pharynx and larynx (Figure 11)
- The laryngoscope is held in the left hand
- With the right hand open the mouth using the scissor maneuver: The thumb on the lower teeth, and the index on the upper teeth
- The laryngoscope is inserted on the right side on the patient's mouth, deflecting the tongue to the left of the blade
- Slide the blade slowly, centrally and forward until you see the epiglottis, and the vocal cords
- With the built-in curve directed anteriorly, attempt to follow the direction of the blade, and advance the ET tube 2 to 3 cm past the vocal cords
- Remove the stylet from the ET tube
- At this point, remove the laryngoscope and inflate the cuff
- Attach the end of the ET tube to the oxygen source
- Make sure to hold on to the ET tube until fixed, using a fixation device or a tape
- If the intubation fails, you should immediately oxygenate and ventilate the patient for 3-5 mins before another attempt

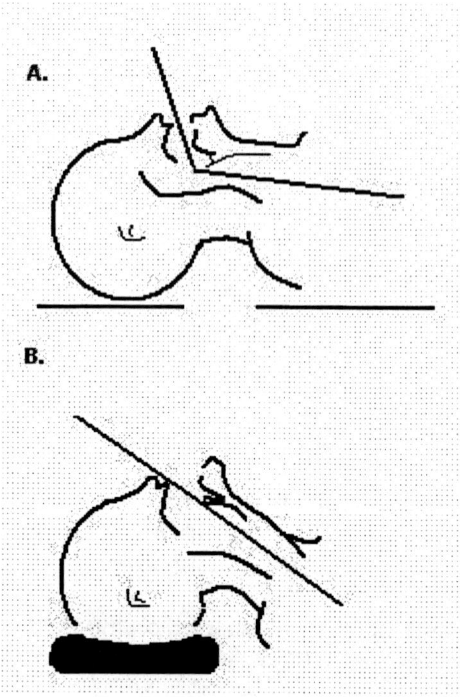


**Figure 11**

Elevation of the patient's head with pads under the occiput and extension of the head at the atlanto-occipital joint (the sniffing position) will allow alignment of the oral, the pharyngeal and the laryngeal axis. This position will permit better visualization of the glottis and the vocal cords.

Cricoid pressure or (Sellick maneuver): By pressing downward on the anterior neck, around the cricoid cartilage, you will occlude the esophagus and minimize the risk of esophageal intubation. It also decreases the risk of gastric distention during mask ventilation.

**Tube Positioning:**

Once the patient is intubated check for the following:

- Bilateral chest movements
- Moisture in the ET tube during exhalation
- A capnograph can be used to detect carbon dioxide in the exhaled air
- Auscultate bilaterally
- Record the length of the tube at the level of patient's teeth; usually it is 23 cm for men and 21 cm for women
- Secure the tube with the fixation device or with the tape
- Order a STAT portable Chest x-Ray; You should visualize the radiopaque indicator embedded in the distal end of the ET tube. The depth of insertion should be assessed, and the tip of the ET tube should be 3 to 4 cm above the carina
- An ET tube placed too low can pass into and ventilate only the right main bronchus, whereas an ET tube placed too high will not provide adequate ventilation

**Complications:**

- Tooth damage and oral soft tissue trauma
- Hypertension, tachycardia and cardiac dysrhythmias
- Aspiration pneumonia
- Injury to the larynx and throat
- Glottic edema
- Intubation of the esophagus rather than the trachea
- Endobronchial intubation
- Accidental extubation
- Tracheomalacia (softening of the cartilaginous rings causing collapse of the trachea on inspiration) and tracheal stenosis (most often at the cuff site) seen in patients with long-term mechanical ventilation

**Ventilator complications:**

- Ventilator failure
- Patient-Ventilator asynchrony (because of inappropriate ventilator settings)
- Hyperinflation, pneumothorax
- Hypotension secondary to decreased venous return
- Pneumomediastinum
- Pneumoperitoneum
- Pulmonary interstitial emphysema
- Subcutaneous emphysema (neck, face)
- Ventilator associated pneumonia
- Nosocomial aspiration
- Oxygen toxicity

**In summary: How to order initial ventilator settings? (AC/SIMV)**

<u>AC mode:</u>

- Set the initial rate at 12 breaths/min
- Set the initial tidal volume at 5-12 ml/kg of body weight
- Set the initial FiO2 at 80-100%
- Assess the need for PEEP
- Set the peak flow at 50 ml/min
- Set the flow triggering (sensitivity)
- Check a blood gas 30 mins later
- If hypoxemia consider adding PEEP
- If CO2 is elevated, consider increasing the ventilator rate or the tidal volume (avoid auto-PEEP)
- Check the peak inspiratory pressure and the plateau pressure
- Choose the setting in collaboration with the respiratory therapist
- Consult Pulmonary/Critical Care

SIMV mode:

- Set the initial rate at 12 breaths/min
- Set the initial tidal volume at 5-12 ml/kg of body weight
- Set the initial FiO2 at 80-100%
- Assess the need for PEEP
- Set the peak flow at 50 ml/min
- Set the flow triggering
- Set the pressure support
- Check a blood gas 30 mins later
- If hypoxemia consider adding PEEP
- If CO2 is elevated, consider increasing the ventilator rate or the tidal volume (avoid auto-PEEP)
- Check the peak inspiratory pressure and the plateau pressure
- Choose the setting in collaboration with the respiratory therapist
- Consult Pulmonary/Critical Care

## 10. Respiratory Monitoring

- Look at the patient, look for any restlessness, assess the mental status, the level of consciousness, assess any use of accessory muscles, the chest and abdominal motion, the skin coloration, the presence of stridor, the presence of excessive secretions or vomitus
- Monitor respiratory rate, heart rate and pulse oximetry
- Transfer the patient to an appropriately monitored setting

**Pulse Oximetry:**

The Pulse Oximetry is a non-invasive method of monitoring the percentage of the hemoglobin saturation. It is linked to a computerized monitor.

The pulse oximeter is a probe which can be attached to the patient's finger, bridge of the nose, ear lobe, or forehead. The monitor will display the $SpO_2$, and the heart rate. It is indicated whenever you want to monitor the oxygen status of the patient.

The Pulse Oximeter is not a substitute for the Arterial Blood Gas (ABG), but can be an early alert to inadequate oxygenation (decrease of the arterial oxyhemoglobin saturation). See oxygen saturation curve.

Few factors that might affect the reading:

- Situations where there is vasoconstriction (eg, cold, hypovolemia)
- Nail polish, artificial nails
- Situations where there is dysfunctional hemoglobin (eg, anemia, methemoglobinemia, carboxyhemoglobin)
- External factors (eg, bright room lighting)

**Capnometry:**

The **Capnometry** is a non-invasive method of monitoring the patient's carbon dioxide. It is primarily used to monitor the patient's ventilatory status.

The **Capnography** is the measurement of the carbon dioxide and the display of the carbon dioxide waveform.

This device displays the amount of carbon dioxide at the end of a tidal breath ($PetCO_2$), providing an estimate of the alveolar $CO_2$ ($PACO_2$) which is a reflection of the arterial $CO_2$ ($PaCO_2$)

$PetCO_2=PACO_2=PaCO_2$ (Because $CO_2$ is rarely diffusion limited)

Clinical uses:

- Patient monitoring in the ICU
- Patient monitoring during general anesthesia
- Prediction of non survival from cardiac arrest (low $PetCO_2$ during CPR has bad outcome)
- To confirm proper endotracheal intubation placement: It should be used as a complement and not as a substitute to the other clinical parameters which are bilateral chest movements, moisture in the ET tube during exhalation and bilateral breath sounds on auscultation

**Table 3**

Common causes of end-tidal carbon dioxide tension (PETCO2) alterations

| Decrease in PETCO2 | Increase in PETCO2 | Absent PETCO2 |
|---|---|---|
| • Cardiac arrest<br>• Decrease cardiac output<br>• Apnea<br>• Hyperventilation<br>• Pulmonary embolism<br>• Leak around endotracheal tube cuff (decreased exhaled gas) | • Increase cardiac output<br>• Hypoventilation<br>• Increase in CO2 production<br>• Fever | • Esophageal intubation |

PETCO2: End-tidal carbon dioxide pressure.

**Table 4**

Example of ventilator checking list

| Date - Time - Therapist | | | | | | | |
|---|---|---|---|---|---|---|---|
| **ABG** | pH | | | | | | |
| | PaCO2 | | | | | | |
| | PaO2 | | | | | | |
| | HCO3 | | | | | | |
| | SpO2 | | | | | | |
| | | | | | | | |
| **Vent settings** | Mode | | | | | | |
| | VT | | | | | | |
| | Rate | | | | | | |
| | FiO2 | | | | | | |
| | VE | | | | | | |
| | | | | | | | |
| **Patient efforts** | VT – rate | | | | | | |
| | Minute ventilation | | | | | | |
| | Neg insp press | | | | | | |
| | | | | | | | |
| **Alarms** | Sensitivity | | | | | | |
| | High insp press | | | | | | |
| | Low insp press | | | | | | |
| | High volume | | | | | | |
| | Low volume | | | | | | |
| | | | | | | | |
| **Dynamics** | PEEP | | | | | | |
| | CPAP | | | | | | |
| | Press sup | | | | | | |
| | Peak flow | | | | | | |
| | Peak press | | | | | | |
| | Plateau press | | | | | | |
| | Static compliance | | | | | | |
| | | | | | | | |
| ET tube | Cuff pressure | | | | | | |
| | | | | | | | |
| Treatment time | | | | | | | |

## 11. Supportive Care

**Suctioning:**

Patients on ventilatory support might need frequent suctioning. Sterile technique has to be used to minimize the risk of infection.

Perform suctioning only when it is needed, (not on a routine basis):

- Direct visualization of the secretions
- Rhonchi on auscultation
- Unexplained increase of peak inspiratory pressure

Oxygenate the patient with 100% oxygen between each suction.

Suctioning can cause:

- Vagal stimulation
- Hypoxia
- Increase intracranial pressure
- Hypertension
- Cardiac dysrhythmias
- Suction trauma

**Chest physiotherapy:**

The aim of chest physiotherapy is to mobilize retained secretions and to prevent further accumulation. There are several techniques:

- The postural drainage: Putting the patient on different positions to promote mobilization of the secretions
- The use of vibration and percussion techniques, in conjunction with the postural drainage

**Nutritional support:**

- Consult nutrition
- Enteral tube feeding, when possible is preferred to parenteral nutrition, it helps maintain the gut integrity
- Use parenteral nutrition in patient with severe gastro-intestinal pathology
- Adequate hydration will help clearance of the pulmonary secretions
- Avoid overfeeding, fluid overload and hyperglycemia
- Monitor the electrolytes
- Stress ulcer prophylaxis
- Aspiration precaution, by routine check of the tube feeding, and elevation of the head from bed at 45° if possible

**DVT prophylaxis:**

Patients receiving ventilatory support are at increasing risk of developing DVT (Deep Vein Thrombosis) and PE (Pulmonary Embolism)

- Use teds stocking or pneumatic compression boots
- If no contra-indications, use either unfractionated heparin or low-molecular-weight heparin

**Pressure ulcer prophylaxis:**

Frequent repositioning and specialty bed help minimize pressure ulcers.

**Sedation:**

It helps decreasing patient's anxiety, oxygen consumption, and improves gas exchange.

Drugs most frequently used:

<u>Benzodiazepines</u> : They inhibit the CNS (central nervous system) by potentiating γ-aminobutyric acid (GABA) receptors.

- Midazolam or Versed: Can be given as IV push every 2-3 mins, acts in 2-3 mins and the duration of action is around 2 hours. Usually used during the intubation, if the patient is agitated.
- Lorazepam or Ativan: Can be given as IV push, acts in 10-15 mins and the duration of action is around 5-10 hours.

<u>Opiates:</u> They are efficient drugs in alleviating the sense of air hunger
- Morphine: Is the commonly used drug, it can be given IV, and the duration of action is 1 to 4 hours.

<u>Propofol or Diprivan:</u> It acts in 1-5 mins and can be given as a continuous infusion because of its extremely short duration of action.
Watch for apnea, hypotension, and bradycardia.

| |
|---|
| Regular and frequent assessments for need of sedation should be performed.<br>Efforts should be made not to over sedate the patients but to help them overcome their anxiety. |

**Patient transportation:**

Frequently, the ventilated patient must undergo diagnostic testing that cannot be done at bedside. Such situation may require the patient to be transported.

Equipments accompanying a critically ill patient should include at minimum:

- Cardiac monitor and defibrillator
- Bag-mask and oxygen tank
- Pulse oximeter
- Blood pressure cuff
- Resuscitation drugs

Make sure that all IVs and tubings are secured. Assign one person to secure the airway and provide ventilation. Keep a close eye on the portable monitor and the pulse oximeter.

**Communicate with the patient:**

The patient should be offered a means of communication with the hospital staff or with the family or friends.
Depending on the patient's condition, the use of letter, word, phrase or image boards can allow a means of communication.

## 12. Weaning

Weaning is the gradual discontinuation of the ventilatory support. Different techniques exist. And the most frequently used are:

1. T-Piece weaning, a spontaneous breathing trial
2. Continuous Positive Airway Pressure (CPAP) with the ventilator rate turned to zero
3. Synchronous Intermittent Mandatory Ventilation (SIMV) with gradual decrease in the ventilator rate

Do not begin the weaning process until the following criteria are met:

- The initial offending process that required intubation is reversed or stabilized
- Avoid excessive sedation
- Vital signs are stable, with improvement of the respiratory drive
- Adequate cardiovascular function, with minimal pressors or inotropes
- Positive gag reflex
- Adequate oxygenation ($FiO_2$ of 40% or less, and $PaO_2$ > 60 mm Hg)
- PEEP of 5 cm $H_2O$ or less
- Correct any Acid-Base abnormalities
- Correct any electrolyte disorders, especially the calcium, magnesium, phosphate and potassium

**1. T-Piece trial:**

Place the patient in the upright or semi-upright position and explain to the patient that you would like to assess if he or she will be able to breathe through the endotracheal tube without the help of the ventilator.

- Disconnect the ET tube from the ventilator and attach it to a source of oxygen
- A low level of CPAP (5 cm H2O) or a low level of PSV (5 cm H2O) can be added
- Watch for 5 to 10 mins
- If the patient is tolerating, continue 30 minutes to 2 hours
- Monitor the RR, HR, oxygen saturation, diaphoresis
- If the patient continues to tolerate the T-Piece, extubate
- Otherwise rest the patient for 24 hours on A/C mode
- Reassess the patient
- Some physicians prefer alternating T-Piece with ventilatory support and increasing each time the length of T-Piece
- Drawback: Increased resistance
- Work in collaboration with the respiratory therapist

**2. Continuous Positive Airway Pressure (CPAP):**

The patient is spontaneously breathing, with a predetermined positive pressure.
The level of the pressure is gradually decreased as tolerated.
When the pressure reaches 5 cm $H_2O$, the patient is extubated.

**3. Synchronous Intermittent Mandatory Ventilation (SIMV):**

The mode is switched to SIMV, with daily, gradual decrease in the respiratory rate by 2-4 breaths/min each time.
A pressure support is usually set to help the work of breathing.
If the patient tolerates a rate of 4-6 breaths/min, extubate after checking an ABG.

## 13. Extubation

If the patient has done well during the weaning process, and if the ABG shows adequate oxygenation, prepare to extubate.
It is recommended not to extubate the patient while you are alone in the room, have the respiratory therapist and the nurse present to assist you.

You will need:

- Gloves
- Oxygen
- A rigid section device: Yankauer
- Bag-mask
- Aerosol therapy
- And of course all what you need to re-intubate

The patient needs to be positioned appropriately (elevation of the head from bed at 45°), and the patient's mouth need to be suctioned well to avoid aspiration.
Administer 100 % oxygen for 2-3 mins.
After deflating the cuff, have the patient take a deep inspiration, and gently pull out the ET tube while suctioning. Then put the patient on humidified oxygen.

There are good chances that the patient's respiratory rate as well as the heart rate will increase after extubation. The patient should be monitored closely for air movements, blood pressure and skin color.
Hoarseness occurs in most extubated patients and should subside quickly.
Stridor is a sign of glottic edema; IV steroids and racemic epinephrine can be given with close monitoring.

If no improvement, the patient may need to be reintubated.

| |
|---|
| Patients that have excessive secretions requiring frequent suctioning or patients that are unable to protect their airways (e.g. decreased level of consciousness) should not be extubated. |

Incentive spirometry (IS): (Figure 12)

- After extubation, especially following surgery, the patient can develop hypoventilation, a complication that can lead to atelectasis and pneumonitis

- When properly used, the incentive spirometer reinflates collapsed alveoli and helps mobilize secretions. (the patient exhales normally, then inhales deeply through the mouthpiece until the balls rise to the chamber tops)

- By looking at the balls moving, the patient knows how well he is performing, the device is a **stimulant** or an **incentive** for him/her

- The patient can rest few seconds between breaths, and can repeat the maneuver 5 to 10 times per hour

- Side effects: Dizziness

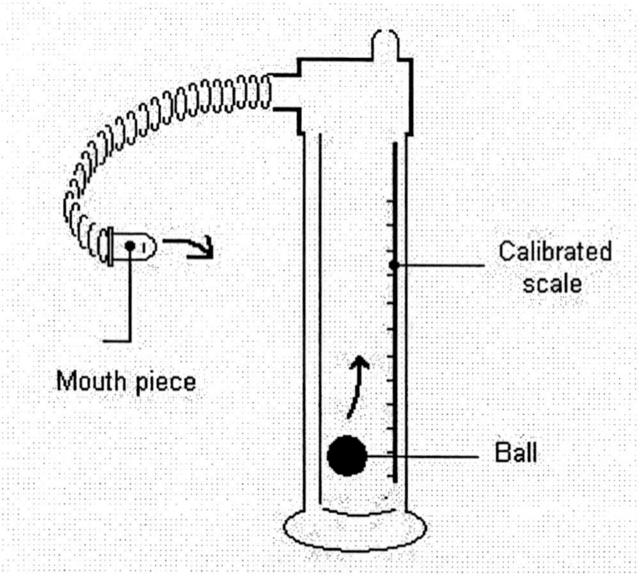


**Figure 12**
Incentive spirometer.

## 14. Tracheostomy

Tracheostomy is a surgical procedure that can be done either in the operating room or at bedside in the ICU. It consists on creating a stoma (an opening) above the suprasternal notch and inserting a tracheostomy tube for long-term mechanical ventilation or airway management.

**Tracheostomy Tubes:** (Figure 13)

A -The outer cannula, to which is attached the cuff

B -The inner cannula, a removable part, for routine cleaning. It comes with a standard 15 mm adapter

C -The obturator with a round tip, used for tube insertion

As with the ET tube, a radiopaque indicator is found in the distal end of the tube.

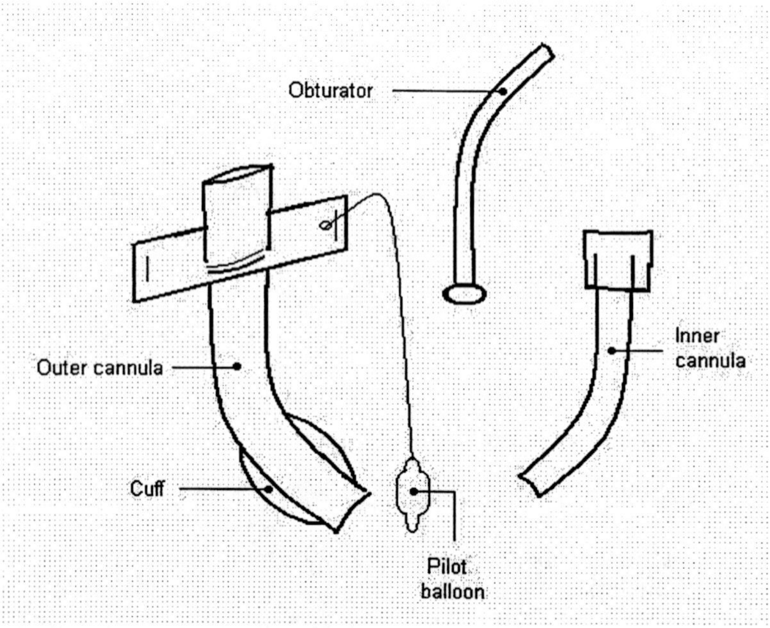


**Figure 13**
Tracheostomy tubes.

**Indications:**

- Prolonged Mechanical Ventilation
- Laryngeal trauma
- Upper airway obstruction
- Inability to protect the airways (eg, decreased level of consciousness, retained secretions requiring ongoing suctioning)

**Early complications:**

1. Apneas
2. Displacement of the tube
3. Aspiration
4. Bleeding
5. Pneumothorax
6. Subcutaneous emphysema
7. Tracheitis, cellulitis

**Late complications:**

1. Aspiration pneumonia
2. Difficulty swallowing
3. Tube obstruction
4. Tracheoesophageal fistula
5. Tracheocutaneous fistula
6. Tracheal stenosis

## 15. Decannulation

Same as in extubation, weaning from tracheostomy requires the following criteria:

- The process that required tracheostomy is resolved
- The patient is able to breathe spontaneously with the tracheostomy cuff deflated, and tube occluded
- Presence of a gag reflex
- The patient is able to clear the secretions
- The secretions are controlled

**Weaning methods:**

There are several weaning methods used:

- Simply removal of the tracheostomy tube
- The use of progressively smaller tubes
- The use of fenestrated tubes
- The use of tracheostomy button

The stoma usually closes within several days.

## 16. Therapeutic Aerosols

Both nebulizers and MDIs with a spacing device can be used to deliver drugs:

**Bronchodilators:**

- Beta-adrenergic agonists
- Anti-cholinergic agents

Beta-adrenergic agonists: Rapid onset of action, induce smooth muscle relaxation via stimulation of the beta-2 adrenergic receptors of the sympathetic nervous system.
Possible side effects: Tachycardia, palpitations and muscle tremors.

Anti-cholinergic agents: Gradual onset of action, induce smooth muscle relaxation via blocking of the cholinergic receptors of the parasympathetic system.

**Steroids:**

Can be administered as aerosols in combination with bronchodilators. They help decrease the airway inflammation and enhance the responsiveness of the beta-adrenergic agonists.
Certain steroids related side effects such as oral thrush may occur.

**Diluents:**

Given to patients with thick secretions, water and saline solutions help in loosening up, and mobilizing thick secretions.

**Mucolytic agents:**

Most commonly used is the N-acetylcysteine (Mucomyst), acts by dissolving the mucus and helps in removing the secretions by coughing, suctioning or postural drainage.

**Anti-infective agents:**

A large variety of anti-infective agents can be used to treat respiratory infections. Appropriate use of antibiotics is based on the infecting agent.

## 17. Chest X-Ray

You should be able to recognize: (Figure 14)

- Skeletal structures: Ribs, spine, clavicle, scapula
- Soft tissue: Axillary, pectoral muscle, neck, diaphragm
- Lung: Parenchyme, fissures
- Trachea
- Carina
- Cardiac silhouette
- Hilium
- Diaphragm
- Costophrenic and costo-cardiac angles
- Ascending aorta
- Descending aorta
- Aortic knob
- Stomach bubble

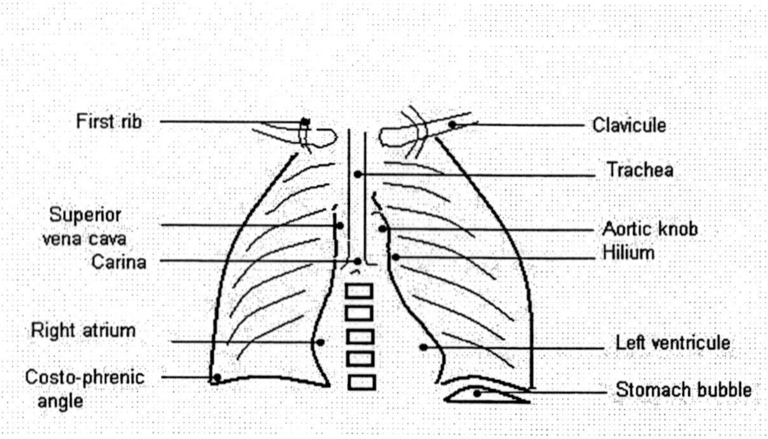


**Figure 14**
Chest X-Ray (PA Projection).

You should be able to recognize:

**Alveolar consolidations:**

Can present as fluffy 3-5 mm nodules, or localized to diffuse densities. They are frequently associated with air bronchogram (aerated airways in consolidated area of the lung).

**Pleural effusion:**

Can be minimal and presents as a blunting of the costophrenic angle or be moderate to severe with complete opacification (increase density) of the hemithorax with shifting of the mediastinum to the contralateral side.

**Pneumothorax:**

Look for area of the lung that may seem blacker (radiolucent) than the other, can be minimal to severe with collapsing of the lung and shifting of the mediastinum to the contralateral side.

**Alveolar edema:**

Look for the butterfly or bat-wing appearance on each side of the cardiac silhouette and the cephalization of flow.

**Interstitial pulmonary edema:**

There is linear or reticular pattern with increase in the interstitial pulmonary marking.

**Emphysema:**

There is hyperinflation of the lungs, with presence of bullae, hyperlucency, flattened hemi-diaphragms and increase retrosternal air space.

**Asthma:**

There is hyperinflation of the lungs, flattened hemi-diaphragms, and increase retrosternal air space.

**Bronchiectasis:**

Look for parallel lines which correspond to dilated and thickened bronchi.

**Cardiomegaly:**

On the postero-anterior view (PA), the thoracic cage should be at least twice as wide as the heart silhouette.
The cardiothoracic ratio is the widest diameter of the cardiac silhouette over the widest diameter of the thoracic cage. The ratio should be less than 0.5.
If the cardiothoracic ratio > 0.5, suspect cardiomegaly.

**Left atrium enlargement:**

There is lateral extension and double density along the right heart border.

**Right atrium enlargement:**

It is difficult to assess on a plain film, there is increase convexity of the lower right heart border on the PA view.

**Left ventricular enlargement:**

Boot-shape of the lower left border of the heart. There is left, downward displacement of the apex.

**Right ventricular enlargement:**

There is rounding and elevation of the cardiac apex and obliteration of the retrosternal space on the lateral view.

## Normal Values

Alveolar ventilation = 4.2L/min
FiO2 on room air = 21%
FRC = 2.40 L
Hb = 14 g/dL in women and 16 g/dL in men
PACO2 = 40 torr
PaCO2 = 46 torr
PAO2 = 100 mmHg
PaO2 = [75-100] mm Hg
PB = 760 mm Hg (sea level)
PH2O = 47 mm Hg at 37˚C
RR = 12/min
RV = 1.20 L
SaO2 = 97%
TLC = 6 L
VC = 4.80 L
VE = 6 L/min
VD = 150 ml
VT = 500 ml

## Glossary

ARDS: Acute respiratory distress syndrome
BiPAP: Bilevel positive airway pressure
CMV: Controlled mode ventilation
CNS: Central nervous system
CPAP: Continuous positive airway pressure
DPG,2,3: 2,3,diphosphoglycerate
EPAP: End-expiratory positive airway pressure
ET-tube: Endotracheal tube
FEV: Forced expiratory volume
FRC: Functional residual capacity
FiO2: Fraction of inspired oxygen concentration
FRC: Functional residual capacity
Hb: Hemoglobin
HR: Heart rate
IMV: Intermittent mandatory ventilation
IPAP: Inspiratory positive airway pressure
I:E : Inspiratory/Expiratory ratio
MDIs: Metered-dose inhalers
O2: Oxygen
PaCO2: Partial pressure of carbon dioxide in arterial blood
PaO2: Partial pressure of oxygen in arterial blood
PAO2: Partial pressure of oxygen in alveolar gas
PCO2: Partial pressure of carbon dioxide
PB: Barometric pressure
PEEP: Positive end-expiratory pressure
PH2O: Partial pressure of water
PIP: Peak inspiratory pressure
PSV: Pressure support ventilation
RR: Respiratory rate
SaO2: Arterial oxygen saturation
SIMV: Synchronized intermittent mandatory ventilation
SOB: Shortness of breath
SpO2: Functional SaO2 as depicted by the pulse oximeter
VA: Alveolar ventilation
VD : Dead space
VE : Minute ventilation
V/Q : Ventilation perfusion ratio
VT : Tidal volume

## Suggested Readings

Following is a list of suggested readings used repeatedly as references for the creation of this manual.


1. Balk RA: The technique of orotracheal intubation: J Crit Illness. 1997;12:316-323
2. Egan DF, Scanlan CL, Wilkins RL, Stoller JK. Egan's Fundamentals of Respiratory Care. Mosby, 1999
3. Ganong WF. Review of Medical Physiology, Lange Medical Book, 2001
4. Hyatt RE, Scanlon PD, Nakamura M. Interpretation of Pulmonary Function tests. A Practical Guide. Philadelphia: LW&W, 2003
5. Heffner JE: The technique of tracheotomy and cricothyroidotomy. J Crit Illness. 1995;10:561-568
6. Kress JP, O'Connor MF, Pohlman AS and Hall JB: Sedating critically ill ventilated patients: A pharmacologic primer. J Crit Illness 1997;12:287-299
7. Legrand TS, Peters JI: Pulse Oximetry: Advantages and pitfalls. J Respir Dis. 1999; 20:195-206
8. MacIntyre NR, Branson RD. Mechanical Ventilation. Philadelphia: WB Saunders, 2001
9. Miller WT, Miller WT Jr. Field Guide to the Chest X-Ray. Baltimore: LW&W, 1999
10. Rodriguez RM, Light RW: Pulse Oximetry in the ICU: Uses, benefits, limitations. J Crit Illness. 1998;13:247-252
11. Rodriguez RM, Light RW: Capnography in the ICU. J Crit Illness. 1998;13:372-378
12. Society of Critical Care Medicine (2002). Fundamental Critical Care Support.
13. Scoggin CH: The technique of weaning from mechanical ventilation. J Crit Illness. 1986;1:59-69

## NOTES

# NOTES

NOTES

NOTES

## NOTES

NOTES

NOTES

NOTES

ISBN 141203119-2

9 781412 031196